LOGGING LONG AGO

Mary L. Martin,
James Edward Thompson, & Tina Skinner

STRAUGHAN'S WORLD FAMOUS LOG HOUSE - FROM CALIFORNIA'S REDWOOD EMPIRE

The log-roller's house

Felling Fir Trees in Oregon

Schiffer Publishing Ltd

4880 Lower Valley Road   Atglen, Pennsylvania 19310

Published by Schiffer Publishing Ltd.
4880 Lower Valley Road
Atglen, PA 19310
Phone: (610) 593-1777; Fax: (610) 593-2002
E-mail: Info@schifferbooks.com

Designed by Mark David Bowyer
Type set in Berylium / Souvenir Lt BT

ISBN: 978-0-7643-2619-6
Printed in China

For the largest selection of fine reference books on this and related subjects, please visit our web site at www.schifferbooks.com
We are always looking for people to write books on new and related subjects. If you have an idea for a book please contact us at the above address.

This book may be purchased from the publisher.
Include $3.95 for shipping.
Please try your bookstore first.
You may write for a free catalog.

In Europe, Schiffer books are distributed by
Bushwood Books
6 Marksbury Ave.
Kew Gardens
Surrey TW9 4JF England
Phone: 44 (0) 20 8392-8585; Fax: 44 (0) 20 8392-9876
E-mail: info@bushwoodbooks.co.uk
Website: www.bushwoodbooks.co.uk
Free postage in the U.K., Europe; air mail at cost.

Contents

Introduction

As America entered the twentieth century, she relied heavily on her prime resource—wood—and the logging industry was in its glory days. The production of lumber reached new heights as a result of mechanical innovations. The images on these historic postcards capture an important part of America's manufacturing heritage—the lumber industry in its heyday from the 1900s through the early 1980s.

From all over the world, men came to do the grueling work of chopping trees and sawing wood in the forests of America. The United States became the hub of the logging industry as evident by the photographic depictions of the slow and meticulous process of harvesting lumber, from cutting down trees to hauling them from the woods to create great flotillas of lumber. This lumber was then to be transported to mills and loaded onto ships for transport to the nation's, and the world's, growing cities.

The gigantic redwoods of California and Oregon were the most challenging for loggers. For days, lumbermen hewed doggedly to fell these huge trees. Living in primitive camps, teams of men hacked tirelessly with axes and saws. Oxen and horses dragged the logs from the deep woods to waiting wagons and sleds that hauled the timber to sawmills. In addition, the mighty waterways provided cheap transportation for lumber companies: river floats. Postcards from the first two decades of the twentieth century illustrate the harsh life that accompanied the arduous work of this period in logging's history.

The technological advances of the late nineteenth and early twentieth centuries allowed lumber companies to expand their operations dramatically to keep pace with the growing demand for wood. Old donkey engines, trains, and ships played an integral part in transporting more timber to sawmills. Also known as vertical gypsies, old donkeys contained huge steam engines and long cable lines, dragging logs from the woods faster than oxen and horses could. In addition to hauling wood to mills, trains and ships carried wood to new markets in the United States and throughout the world. This visual tour also takes us inside the sawmills where we see the increasing mechanization of lumber production that accompanied the expansion of the logging industry.

As the twentieth century continued, lumber companies expanded their operations by adding other facilities to their mills. Making use of every bit of wood, these companies built planing mills that produced large quantities of lumber. Paper was made from huge piles of pulp. A few mills in California concentrated solely on producing shingles for the construction industry. Lumberyards became synonymous with the sawmill. The need for lumber soared when America entered World War II. The military needed wood for gunstocks, barracks, stretchers, paper, bridges, and vehicles. The demand for wood products was at its highest.

After World War II, the military sold old, surplus trucks to the lumber companies. This good fortune initiated yet a new phase to the logging industry. The massive trucks contained powerful engines that allowed them to travel on old logging roads once reserved for oxen and horses. Logging companies immediately recognized the advantage of such vehicles and began using trucks to haul logs from the deep woods to their mills. Small trucks to huge tractor-trailers became familiar sights in logging country. This period of development is illustrated wonderfully with color postcards from World War II through to the 1980s.

The logging industry underwent many changes in its history. The romance of the woodsman's life gave way to lumber camps and company towns that provided security for a few generations. Declining demand for wood and environmental conservation impacted the logging industry in the second half of the twentieth century. As a result, mills began to close and lumber companies went out of business. In spite of such setbacks, the logging industry survives today with new innovations such as chainsaws, cranes, and loaders. Once devastated by logging, the land is slowly being restored through the advent of tree farming. A new chapter in logging's history is being written. These colorful, hand-tinted, black and white postcards trace the early chapters of this exciting story, bringing logging in the early twentieth century to life.

Historic Images Through Postcards

Postcards are said to be the most popular collectible history has ever known. The urge to horde them sprang up with the birth of this means of communication at the turn of the twentieth century, and has endured great changes in the printing industry. Today, postcard shows take place every weekend somewhere in the country, or the world, and millions of pieces of ephemera lie in wait for those who collect obscure topics or town views.

Postcards once served as the email of their day. They were the fastest, most popular means of communication beginning in the 1890s in the United States. These timely cards provided a way to send visual scenes through the mail along with brief messages—a way to enchant friends and family with the places travelers visited, to send local scenes, or to share favorite topics of imagery. They even provided the latest breaking news, as images of fires, floods, shipwrecks, and festivals were often available in postcard form within hours of an event. Moreover, mail was delivered to most urban homes in the United States at least twice a day. So someone might send a morning postcard inviting a friend to dinner that evening, and receive an RSVP in time to shop for food.

The messages shared and the beautiful scenes combined to create the timeless appeal of postcards as a collectible. Most importantly, history is recorded by the pictures of the times, moments in time reflecting an alluring past.

Dating Postcards

Pioneer Era (1893-1898): Most pioneer cards in today's collections begin with cards placed on sale at the Columbian Exposition in Chicago on May 1, 1893. These were illustrations on government-printed postal cards and privately-printed souvenir cards. The government cards had the printed one-cent stamp, while souvenir cards required a two-cent adhesive postage stamp to be applied. Writing was not permitted on the address side of the card.

Private Mailing Card Era (1898-1901): On May 19, 1898, private printers were granted permission, by an act of Congress, to print and sell cards that bore the inscription, "Private Mailing Card." A one-cent adhesive stamp was required. A dozen or more American printers began to take postcards seriously. Writing was still not permitted on the back.

Post Card Era—Undivided Back (1901-1907): New U. S. postal regulations on December 24, 1901 stipulated that the words "Post Card" should be printed at the top of the address side of privately printed cards. Government-issued cards were to be designated as "Postal Cards." Writing was still not permitted on the address side. In this era, private citizens began to take black and white photographs and have them printed on paper with postcard backs.

Example of a postcard with an undivided back. Senders could only write the address on this side of the card. Any message needed to be written on the front of the card along with the picture.

Early Divided Back Era (1907-1914): Postcards with a divided back were permitted in Britain in 1902, but not in the United States until March 1, 1907. The address was to be written on the right side; the left side was for writing messages. Many millions of cards were published in this era. Up to this point, most postcards were printed in Germany, which was far ahead of the United States in the use of lithographic processes. With the advent of World War I, the supply of postcards for American consumption switched from Germany to England and the United States.

White Border Era (1915-1930): Most U. S. postcards were printed during this period. To save ink, publishers left a clear border around the view, thus these postcards are referred to as "White Border" cards. The relatively high cost of labor, along with inexperience and changes in public taste, resulted in the production of poor quality cards during this period. Furthermore, strong competition in a narrowing market caused many publishers to go out of business.

Linen Era (1930-1944): New printing processes allowed printing on postcards with high rag content that created a textured finish. These cheap cards allowed the use of gaudy dyes for coloring.

Photochrome Era (1945 to date): "Chrome" postcards began to dominate the scene soon after the Union Oil Company placed them in its western service stations in 1939. Mike Roberts pioneered with his "WESCO" cards soon after World War II. Three-dimensional postcards also appeared in this era.

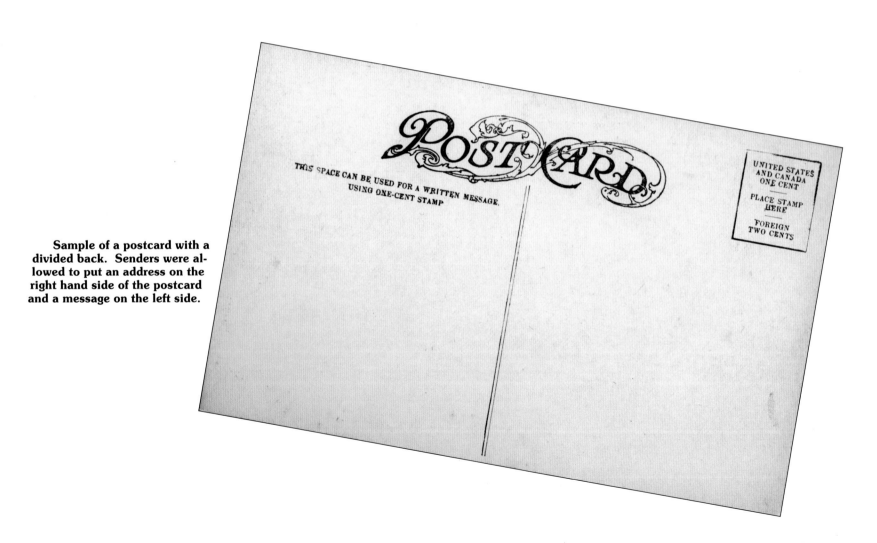

Sample of a postcard with a divided back. Senders were allowed to put an address on the right hand side of the postcard and a message on the left side.

Marveling at Size

S ince the first axes cut into the mighty redwoods, firs, cedars, and hemlocks of the Northwest, men have been in awe of these gigantic trees. The comparison of people to the enormous trees is astounding, as these scenes resemble Liliputians surrounding Gulliver. Everything about these huge trees was great, including the trunks and pine cones.

Workers surround a Boole Tree, one of the ten largest Sequoias. It is over 100 feet in circumference. The huge trunk provides some natural platforms on which to stand.

Cancelled 1914, $4-6

From coast to coast, pioneers to American soils found virgin forest towered over by massive trees. Here, three friends are pictured measuring the girth of a massive white pine in the Allegheny National Forest of Pennsylvania.

Circa 1920s, $4-6

An Oregon Fir Tree, 25 feet in diameter at base, 300 feet high

A horse and rider pose calmly within a cutout section of a large trunk.

Cancelled 1908, $4-6

Standing upon a felled California Redwood, a logging team illustrates their accomplishment for a photographer.

Cancelled 1909, $4-6

A logging team on a felled California Giant 26 feet in Diameter

Standing among massive tree roots, a group of men join hands at the base of an Oregon fir, twenty-five feet in diameter.

Circa 1910, $4-6

A troop of cavalry measures up against a fallen California Redwood.

Circa 1910, $7-9

Redwood Tree "Wawona," California

A tunneled-out section of a huge redwood provided an interesting ride.

Circa 1907, $4-6

It's the end of the line for this stagecoach ride upon a fallen tree in Mariposa Grove, California.

Cancelled 1908, $4-6

390 - A Fallen Monarch, California Big Trees.

Posing on an impressive perch in California.

Circa 1910, $4-6

A HORSE ARCHWAY.

Everything about this "California Sugar Pine" is huge, even the three piles of pinecones.

Circa 1909, $4-6

A card published in England extols the giant trees of the Pacific Coast of the United States as "one of the world's greatest curiosities. They occur in the States of Oregon, Washington, and California.... Some of these trees have no branches nearer the ground than 100 feet, and many of the branches are fully six feet in diameter."

Circa, 1907, $4-6

202. A Giant of the California Forest, cut by the Woodsmen

One section of a huge California tree was plenty of work for these woodsmen and their axes.

Cancelled 1908, $4-6

A little girl poses by a redwood log, secured and ready to be transported to a mill.

Circa 1930, $5-7

A Stick of California Stove Wood.

There was probably plenty of "California stove wood" from this "stick." Note the thickness of the bark as revealed by the exposed layer.

Circa 1912, $4-6

CALIFORNIA LOGGING SCENE   4593

The ends of these logs show the hard work involved in cutting down these huge trees—the angled axe chops were created to gain admittance for a two-man saw.

Cancelled 1938, $4-6

A Twelve Foot Spruce Log - 12,720 Board feet.   W 127

The giant California trees did not come down easy as shown by the tears in the wood grain. This 12-foot log was estimated to produce 12,720 board feet.

Circa 1928, $4-6

13

"Boole" The largest tree in the World cir. 109 ff., Fresno Co. California.

Boasting a circumference of 109 feet, this tree plays host to a party near Fresno, California. Now part of Sequoia National Forest, the tree is still a popular attraction, and deemed the largest tree in the United States.

Circa 1908, $4-6

Members of a lumber camp crowd a fallen California Redwood and pose for a photograph. Virgin redwood forests once covered 3,000 square miles of land extending 500 miles along the West Coast starting south of Monterey, California, and extending just north of the Oregon border.

Circa 1907, $4-6

Two great sections of tree were hollowed to create habitat and tourist lure, surrounded by generous pickets of spare lumber.

Circa 1907, $5-7

The entire logging camp poses in a card entitled, "Leisure hour in a Western logging camp."

Cancelled 1912, $5-7

A cross-section of a tree displays its age in relation to world events for visitors to Muir Woods National Monument near San Francisco, a rare stand of preserved virgin redwood forest. The park was created to honor John Muir, a naturalist, explorer, and advocate for forest conservation.

Circa 1930, $5-7

Lying on the ground for 2,500 years—adequate time for a young redwood to grow up around it—this enormous redwood stump is said to be still sound. Redwood is remarkably immune to rot and insect infestation, and a forest of live redwood can withstand even the hottest forest fire and survive.

Circa 1930, $4-5

Located on a campground near Cummings, California, and variously named throughout its history, this "small section" of a log served as restroom accommodations and the subject of many postcards throughout its history. Here it is as it stood when the site was named Grundy's in northern California. The structure was torn down in the early 1960s.

Circa 1962, $6-10

SPRUCE LOG AT PALMER PARK, DETROIT, MICH.

Named both the "Mark Twain Stump" and the "Big Stump," this tourist attraction has been drawing curiosity seekers to Kings Canyon National Park in California since it was cut down in 1891. The 1,700-year-old tree took 13 days to fell, being 26-feet wide. Because sequoia takes decades to decay, piles of sawdust from the cut still remain on the site. Sections of the tree were sent to the American Museum of Natural History and to the British Museum in London, England.

Circa 1970, $4-6

A hollowed-out spruce log was set on display in Palmer Park in Detroit, with the accommodations that included a writing desk in one room and an animal cage at the other end.

Cancelled 1917, $4-6

A small undercut is not the most ideal place to rest or sleep, but in a pinch, it'll do as it does for these two confident workers who pose within a twelve-foot-wide undercut on a Douglas fir in Washington state.

Circa 1928, $5-7

Large logs were used as support columns within the Forestry Building.

Circa 1907, $4-6

7467. Forestry Building, Portland, Ore.

INTERIOR OF FORESTRY BUILDING, LEWIS AND CLARK MEMORIAL, PORTLAND, ORE.

The impressive Forestry Building was built in Portland, Oregon, for the 1905 Lewis and Clark Centennial American Pacific Exposition. Built primarily of Douglas fir, with four cedar columns at the entrance, the building was said to contain more than one million feet of board lumber. Unfortunately, this landmark building burned to the ground in 1964.

Circa 1907, $4-6

680 A FALLEN GIANT, OREGON

The center of this hollow log was adequate for hiding a horse and rider.

Cancelled 1910, $4-6

Cut from a redwood tree, a thirty-three-foot-long log was transformed into a unique log house. Cut on the property of the Hammond Lumber Company, the 1,900-year-old tree was ample enough to create four such logs.

Circa 1948, $4-6

STRAUGHAN'S WORLD FAMOUS LOG HOUSE - FROM CALIFORNIA'S REDWOOD EMPIRE

19

Lumber Camps

In the early days of the logging industry, the lumber camp was the way of life. Generally located near the source of the industry, the camps were remote from general society, and peopled almost exclusively by men. In a life lived close to the elements, the camp offered primitive conditions to its temporary inhabitants. These simple, mobile camps provided for the most basic needs—a place to return to nightly and consume the massive calories needed to fuel another day's labor, and a dry place to sleep before the next long workday got underway.

Every lumber camp needed a grindstone to sharpen axes and saw blades.

Dated 1905, $6-8

6. LOGGING CAMP, NORTHERN WOODS.

Loggers are pictured seeking shelter in ice-encrusted cabins.

Circa 1910, $4-6

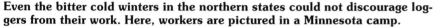
Lumber Camp, Northern Minnesota.

The log-roller's house

Even the bitter cold winters in the northern states could not discourage loggers from their work. Here, workers are pictured in a Minnesota camp.

Cancelled 1908, $6-8

Log rollers or "hand loggers" rest outside their shack. After a tree was cut down, this pair would cut off the branches, enabling the log to be rolled down a hill. This card was produced as a fundraiser by the Reverend W. S. Clairborne of Tennessee who proclaimed, "Practically the only religious influence in the lives of these wild people is supplied by our eight mountain missions, which in many cases have entirely transformed their lives. Will you help me..."

Circa 1920, $6-8

9545. A SOUTHERN LOGGING CAMP.

A day of rest in camp allowed men to wash their clothes or to get a shave or haircut.

Cancelled 1907, $4-5

Lumber Camps along the St. Croix.

Deep snow and freezing temperatures added to the hard life at a lumber camp in northwestern Wisconsin.

Circa 1920, $4-6

Logging Camp, Bridal Veil Falls, Oregon.

Rolling and floating logs in a camp in northern Oregon near its border with the State of Washington.

Dated 1907, $4-6

6914   LOGGING SCENE NEAR MARINETTE AND MENOMINEE.

Despite the harshness of life at a logging camp in winter, nature often painted a serene picture.

Circa 1914, $5-7

Trains eventually replaced oxen and horses for transporting logs to and from camp. This card is labeled a "Modern Logging Camp" in Washington State.

Dated 1907, $5-7

The Lowman & Hanford Co., Seattle

1140. Modern Logging Camp, Washington.

Inside the cook's shanty—where the cook and his helpers prepared three meals a day for over twenty men in camp.

Circa 1930s, $7-9

Cook Shanty.

23

At least three to four men used long, metal-tipped wooden poles to roll and turn a section of log so that the bark could be stripped off and the log hewed or cut further.

Circa 1912, $6-8

Leaving the comfort a warm bunkhouse in New York State, a group of lumberjacks head out into knee-deep snow to carry on their work.

Circa 1907, $6-10

Cutting Down Trees

Day after day, rugged men hacked away at the mighty trees of the forests. The sounds of axes and saws were familiar rhythms to lumberman. The brawny lumbermen swung their axes and saws with raw power. Cutting down trees was a hard day's work requiring teams of men for the task. The image of a man with an axe chopping into a tree is firmly etched in the history of logging.

Cutters work in tandem to saw a felled tree into manageable lengths so that the pair of oxen can haul the lumber back to the camp.

Circa 1915, $8-12

4 DOWNING TIMBER, NORTHERN MINNESOTA.

Brute strength and incredible stamina were needed to bring down the mighty trees.

Circa 1912, $4-6

WESTERN TIMBER SCENE.

Springboards were inserted into trees allowing cutters to "climb" so that they could remove the larger limbs.

Circa 1910, $4-6

Felling a big Redwood, Fresno Co., Calif.

963 Cutting down big Redwoods, in HUMBOLDT COUNTY, California.

150—Giant Fir. One of the Big Trees of the West

These two wiry lumberjacks created springboards and platforms to gain access to this giant redwood.

Circa 1910, $4-5

You can imagine the deafening sound of branches breaking and wood cracking as this huge redwood fell near Fresno.

Cancelled 1911, $4-5

To undercut this giant fir, a small section of outer bark was removed by the wood choppers. Removing the bark made it easier to cut the wood of the interior layers.

Circa 1920s, $4-6

27

Some California Big Trees. A gigantic Redwood Trunk was exhibited at the World's Fair, Chicago

Felling a huge Redwood in the Big Tree Grove, Santa Cruz, Cal. 2911

A bearded worker poses with his axe in an undercutting of a redwood tree felled near Santa Cruz, California.

Cancelled 1923, $4-6

Days were spent cutting this tree and preparing a cross-section of it for transport to the World's Columbian Exposition of 1893 in Chicago. Photos were taken like this one, and another with men, arms extended standing seven abreast at the base of the tree. Still, the folks back at the fair were convinced that it was all a hoax, the sequoia being unimaginably huge.

Circa 1910, $7-10

165 — GIANT FIR TREE

EDWARD H. MITCHELL. PUBLISHER, SAN FRANCISCO.

548 Twenty-six foot saw used for cutting Calif. Redwood tree.

Dual-handled crosscut saws like this twenty-six-foot blade were used by a pair of men to cut down the giant redwood trees.

Circa 1907, $4-6

It was a hard day's work just to carve out an undercutting.

Circa 1907, $4-6

Undercutting a tree was tedious and strenuous work as well as dangerous.

Circa 1920s, $4-5

An Oregon Sapling.

Woodsmen pose for a picture of their hard work.

Cancelled 1910, $4-6

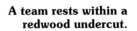

Undercutting a Redwood. 4927

A team rests within a redwood undercut.

Circa 1920s, $4-5

Chopping in the Redwoods   4390A

Felling Fir Trees in Oregon

O 62

PUBLISHED BY E.P. CHARLTON & CO. PORTLAND, OR.

AN OREGON FIR TREE, 9 FEET IN DIAMETER

When using the springboards, woodsmen exercised a combination of balance and tremendous strength to cut down a tree.

Circa 1905, $5-6

A rare image of a topper—one of a select breed of brave men who would work their way to the top of a tree removing the limbs and finally the top of a tree. Only after this work was done would the great tree be felled, minimizing the damage to other trees in the forest during its route to the ground.

Circa 1960s, $4-6

A thick layer of bark had to be removed before these men could get to work felling a fir tree in Oregon.

Circa 1920, $4-5

31

The job of buckers was to cut a downed tree into sections and remove the rest of the branches. Because sections often rolled free after being separated from the rest of the tree, many men were injured.

Circa 1910, $4-5

SAWING A FALLEN TIMBER LUMBER SCENE, WASHINGTON.

Even cutting down a huge tree stump for another log section was a fantastic feat.

Cancelled 1909, $4-5

A Giant Stump of the Pacific North West

IN WASHINGTON'S FORESTS A FALLING TREE
PHOTO BY H. S. WATERS

Another casualty in the deep woods, an action shot catches the demise of a tree in Washington State.

Circa 1907, $4-6

Fir Tree near Everett, Wash.

Workers ham it up for a photographer in the big woods of Washington State.

Circa 1905, $4-6

Springboards provide access to a Washington cedar tree.

Cancelled 1908, $4-6

28. Lumber Jacks, Washington Cedar Trees.

33

176. A GIANT FIR TREE IN A WASHINGTON FOREST.

The double-edged axe posed a danger to the chopper on the back swing. This cedar tree and its undertaker were photographed in South Bend, Washington.

Cancelled 1912, $4-5

Loggers show off by a giant fir in Washington State.

Circa 1917, $4-6

Workers take their lunch on site, probably in one of the eastern forests. This real photo postcard bears no identification other than the "Lunch time" caption a photographer scratched onto the glass negative.

Circa 1915, $7-10

Timber—

DETROIT PHOTOGRAPHIC CO., PUBLISHERS.

7178. FELLING A BIG TREE.

March 1 st 1906

Woodchoppers worked in
pairs to undercut a tree.

Dated 1906, $5-6

A tree falls, presumably in Washington State
where the card was cancelled. On the back the
author writes, "No, I'm not chopping down trees,
but as you can see from this card there are plenty
of them around here."

Cancelled 1949, $7-10

35

Tree cutters pair up for cold-weather work, probably in the woods of Great Lakes region, though this real photo postcard bears no identifying information.

Circa 1905, $4-6

Younger men came to the camps to earn money for their "stake"—starter money to buy a farm or a business. Others came seeking adventure.

Circa 1905, $4-5

Cutters kept two bottles with them while they worked—one was drinking water, the other was oil for the saws.

Cancelled 1909, $4-5

Amidst the Wreckage

The carnage of logging was unavoidable. Felled trees blanketed the floors of great forests like the aftermath of a great battle. Wood chips, broken branches, and endless sawdust lay among the tree stumps. Such destruction was necessary to produce lumber. Clear spaces in the mountains and forests remain today as scars of logging.

1132. Timber Scene, Washington.

The carnage of logging—fallen trees, broken branches, tree stumps, torn ground, and wood chips spewed about.

Cancelled 1959, $4-6

The lumber camps were a melting pot of nationalities with men from various backgrounds and ages, sometimes including ex-soldiers of foreign countries. All had to adapt to difficult living conditions and a work environment fraught with danger.

Circa 1916, $7-9

9544 IN THE SOUTHERN LUMBER DISTRICT.

Trees were cut into uniform lengths of 24, 32, and 40 feet to be hauled to the sawmills.

Cancelled 1910, $4-6

3 WOODSMEN OF NORTHWEST.

The familiar look of the rugged lumbermen—burly men with mustaches and beards dressed in overalls or suspenders.

Circa 1908, $4-5

Logs were put on skids—smaller logs spaced to allow the larger sections to be rolled and moved while the buckers worked.

Cancelled 1909, $4-6

An area of land was cleared for the camps to pile logs ready for the sawmills. Men pose with the long poles used to roll the logs from place to place.

Circa 1904, $6-8

Sections of a California Redwood tree with flat and angled cuts lying amidst many pieces of scrap wood. This photo was taken by A.W. Ericson in Arcata, California.

Cancelled 1910, $6-8

In the California Redwoods, TREE 17 x 19 FEET IN DIAMETER.
From Photo. by A. W. ERICSON, Arcata, California.

Logging in Western Washington.

If logs rolled free, they might tumble half a mile downhill—one of many dangers faced by loggers.

Cancelled 1909, $6-8

Woodsmen stand amidst felled fir in Washington.

Cancelled 1912, $4-6

2147   Felling Timber, Washington.

Hauling Out Timber

Once a tree was felled, the next challenge entailed hauling it out of the forests. The deep woods and rough terrain presented numerous obstacles to this goal, but lumber companies adapted. They built skid roads and used oxen and horses to drag logs from the woods. Later, "donkey engines" with cable lines replaced the mighty beasts of burden.

Logging Train And Town about 1900

Ark Raft Kitchens about 1900

Water floats and rails were two ways of moving massive amounts of lumber. The text on the back of this card, created for the City View Motel in Williamsport, Pennsylvania, commemorates the town's history as the "lumbering capital of the world" in the 1870s.

Circa 1970, $4-6

Lumbering near

Difficult terrain was common to lumber camps, so smaller logs were imbedded in the ground to create "skid roads." These roads were designed to make hauling and traveling easier.

Cancelled 1909, $4-6

After hauling logs to the piles, loggers rolled the logs over long braches or poles onto the huge piles.

Circa 1907, $4-6

Teams of
horses are shown hauling
wagonloads of lumber from the woods
near Superior, Wisconsin.

Dated 1893, $4-6

A team of horses hauls a
string of five-foot-high logs
down a skid road.

Circa 1910, $4-6

Horsemen bragged of their animal's strength and power, and exercised it daily with work in the woods.

Cancelled 1909, $4-6

Steam powered engines and trains enabled more timber to be transported out of the woods quicker than ever.

Cancelled 1914, $4-6

Mechanical logging began with the "vertical gypsy," a machine that contained a single-cylinder engine and a vertical boiler connected to a drive shaft.

Circa 1914, $4-6

The vertical gypsy hauled the larger logs from the forests to where the line horse waited to drag them to the sawmills.

Cancelled 1919, $4-6

265. HAULING BIG LOGS OUT OF THE TIMBER BELT, OREGON.

Sometimes known as a donkey, the vertical gypsy pulled logs to the waiting lumber trains with the help of chains.

Circa 1910, $4-6

44

1162 — Logging the Largest Timber in the World. Redwood.

Wooden chutes were used with the mechanical donkeys to transport logs to the sawmills as well.

Circa 1910, $4-5

Hauling big trees of the Pacific Coast with donkey engines.

Cancelled 1908, $4-5

The engines of vertical gypsies improved with several mechanical advancements—bigger engines allowed the donkey to yard, or haul, more logs as well as bigger logs from the forests.

Cancelled 1908, $4-5

Logging Scene.

The steam engine of the donkey was up and running by 6 a.m. and had to be fed all day.

Circa 1907, $4-5

Several men worked together in the gypsy's operation. One or two men ran the donkey, while one kept the fire going, and still another made sure the logging line was clear.

Cancelled 1911, $4-6

Felling Timber in Washington.

Several men assisted the "donkey puncher"—the man who operated the donkey engine. One man walked the line horse that pulled the cable line into the woods, while another fueled the engine with wood. After a hook tender secured a cable around a log, a whistle punk yanked on the whistle wire to tell the puncher to start the engine and begin pulling the log out. This photograph was taken in 1913 near Yelm, Washington.

Circa 1945, $4-6

Published in England, this card extols the logging by steam in western America: "The logs are dragged out of the forest and to the end of the road where they are loaded on railroad trucks by means of powerful wire rope cableways. The cables are nearly an inch thick and one thousand or more feet in length, and they wind on drums operated by powerful engines, which drag the logs over the roughest kind of road at surprising speed."

Circa 1907, $5-7

STEAM LUMBERING.

Lumber, El Dorado County, Cal.

Piles of lumber were towed on platform cars, ready to be unloaded into the water for a river float. On the back: "The forests of El Dorado County, California, furnish millions of feet of pine lumber for the market annually. This land cleared of large timber may be purchased at $15.00 per acre. It affords a wonderful opportunity for the home seeker with limited capital. Climate, soil and water are unexcelled."

Circa 1907, $4-6

47

A PROSTRATE GIANT.

Also published in England, this card teaches that: "Where the old-time lumberman in other localities depended upon conveying his logs by ice roads in winter, or floating them down streams at spring freshets, the Pacific Coast lumberman has to get along without any such aid, for though the latitude in which the big trees are found is almost the same as that of Labrador, snow and ice are almost unknown."

Circa 1905, $6-7

CLEARING WHEAT LAND.

Another card published for "Their Majesties the King & Queen" was titled, "Clearing Wheat Land. One of the up-to-date features of agricultural work is the clearing of wheat land by means of giant traction engines, ranging in power from 60 to 100 horse-power. In this picture, one of these monster tractors is seen hauling 15,000 feet of logs on 2 trucks down a 17 per cent grade."

Cancelled 1908, $4-6

The remnants of a bygone era: Old donkeys no longer in use lie at rest in Grays Harbor County, Washington. *Photo by John F. McNamara*

Circa 1940s, $4-6

THE WASHINGTON WAY OF LOADING LOGS ON A TRAIN.

Photo by H L Toles

NO. 36. LOADING LOGS IN NORTHERN WOODS.

The next innovation in logging: The old donkey, now sitting on a large platform like a crane, allowed more logs to be loaded onto trains or wagons.

Cancelled 1911, $5-7

Eventually, the old donkeys were given bigger engines in order to perform the more difficult work of loading logs on a train.

Cancelled 1915, $4-6

Logs are unloaded at the sawmill after being hauled on a small railway car. As you can see, the man in the center stands in some danger of being crushed by two logs.

Circa 1907, $4-5

A Happy Christmas

Christmas—when we hear it's name, what memories sweet come thronging;
May you be blest with all that's best, unto this day belonging.

A Washington "Big Stick".

A huge log secured on a typical platform at a Washington lumber mill.

Cancelled 1912, $4-6

LUMBERING NEAR DULUTH, DULUTH, MINN.

NORTHWEST LOGGING SCENE

Tracks were laid to the edge of the forest allowing railroad platforms to haul logs in and out of the woods, eventually replacing the ox and horse.

Circa 1907, $4-6

Large trucks eventually replaced the railroad cars; they were able to go where railroad tracks couldn't.

Circa 1940s, $4-6

LOGGING BY MOTOR TRUCK IN THE NORTHWEST

Logging equipment became just as huge as the trees being felled in the modern logging industry. *Photo by U. S. Forest Service.*

Circa 1955, $4-6

A modern logging scene included the crane-like donkey that pulled logs along a skid road.

Circa 1930, $4-6

Large cranes, bulldozers, and tractor-trailers are some of the massive equipment now employed by the logging industry. This image was taken at the Union Lumber Company in Fort Bragg, California.

Circa 1970s, $4-6

River Floats

A river float was a sight to see with its massive assembly of logs drifting on the water. It was a cheap method of transportation, but fraught with danger. Men jumped and treaded over the logs poking at strays, guiding the boom in order to keep it on course. They battled raging, cold waters, sandbars, winding waterways, and runaway logs. Out of necessity, the river men who drove and worked the floats were full of daring and adventure.

After the logs were dumped into the water, they were sorted by size to be put into booms, the name given to the floating groups of logs.

Circa 1940s, $4-6

Another way to transport logs to the sawmill was the river float, where cut trees were guided by a boat or craft down river or along a coast.

Cancelled 1958, $6-8

Booms were designed to move large piles of logs easily and safely in the water. Here, beached logs are shown by the Gardiner and Randolph Bridge in Maine.

Cancelled 1909, $4-6

Gardiner, Me. Gardiner and Randolph Bridge.

Logs jam at a bridge in Skowhegan, Maine.

Cancelled 1909, $4-6

Skowhegan, Me., M.C.R.R. Bridge.

Logs fill the Kennebec as far as the eye can see in Maine. Each year the river provided transport to more than 150,000 cords of wood, according to text on the back of the card.

Circa 1960s, $4-6

Logs fill the river running through Rumford Falls, Maine.

Cancelled 1906, $4-6

Logs are eased over the dam at Pontook Dam in Maine.

Circa 1912, $4-6

Winnepesaukee River &
Lake Winnisquam,
Laconia, N. H.

A bag boom did not contain a particular arrangement of logs. The group was surrounded by just boomsticks, logs that were secured together with chains, which kept the logs from floating apart on the water. This image was taken in Laconia, New Hampshire.

Circa 1907, $6-8

Logs lashed together form a barrier to keep a docking area free of logs being sent down the Connecticut River in Holyoke, Massachusetts.

Cancelled 1911, $4-5

The Boom on Connecticut River, Holyoke, Mass.

U. S. Series 119/5. Timber Raft, Williamsport, Schute.

This image shows a raft created from logs and used to float lumbermen and a hauling team down the Connecticut River.

Cancelled 1906, $6-7

Workers create a timber raft. A major problem of the large rafts was that they moved slowly and required calm water conditions to ensure complete delivery.

Circa 1910, $4-6

An image by one of America's pioneer photographers, H. H. Bennett, pictures lumber raft running at the Dells in Wisconsin. The rafts had to be disassembled to negotiate the winding channel at Munger's Mill, near the site of the present dam. The eight-foot drop presented many hazards to the raftsmen.

Circa 1960s, $4-6

LOG DRIVING IN THE ADIRONDACK MOUNTAINS, N. Y.

Directing logs down raging waters was not an easy task. Often, the boom broke apart or had to be disassembled to go through narrow straights. Here, men are pictured helping to guide logs down a narrow chute in the Adirondack Mountains of New York. The waterfall on the other side could easily splinter their valuable cargo.

Cancelled 1916, $4-6

A contemporary photochrome card shows men working the annual spring log drive on the White River in Ontario.

Circa 1960s, $4-6

SCENE NEAR WINDHAM, N.Y.

How did he get there? The answer is, he crossed the logs. Boom men were nimble dancers who stood on boomsticks and used long pike poles to keep logs from drifting away and to help the booms navigate the rivers. They also dared to cross on un-tethered logs, running the risk of slipping between rolling logs and being crushed or trapped below the surface of the water.

Circa 1920s, $4-5

Not only did loggers have to contend with unpredictable river conditions, they had to deal with the tidewaters in coastal areas as well. Here, a long and carefully lashed boom floats on the Bayou Teche in Louisiana.

Circa 1914, $5-6

A paddle boat is used to guide booms along the Mississippi River.

Cancelled 1909, $4-6

A log landing in Lake Newman, Florida. A daring soul stands on one of the furthermost logs within the water.

Circa 1910, $4-6

Another difficulty encountered during the river floats was navigating the narrow channel skillfully in order to prevent booms from breaking up or getting stuck along the banks. Here, two workers, one in a boat, the other perched atop a log, negotiate the Suwannee River in Florida.

Cancelled 1921, $4-5

A great harvest is fed to a mill in Duluth, Minnesota.

Circa 1905, $5-6

A great mound of logs lies on the banks of the Chippewa River in Wisconsin.

Cancelled 1912, $4-5

Logging, Chippewa Falls, Wis.

Two loggers ride a log in the waters of Chippewa Falls, Wisconsin.

Cancelled 1917, $4-6

LOG JAM, DELLS POND, EAU CLAIRE, WIS.

A log jam in Dells Pond, located in Eau Claire County, near Augusts in northwest Wisconsin.

Circa 1910, $4-6

Another image by photographer Henry Hamilton Bennett shows workers gathering parts of their raft to reassemble after the falls, before continuing on down the wider stretches of the lower Wisconsin River and, possibly, on to the Mississippi.

Circa 1960s, $4-6

Workers prepare logs for their journey on Wolf River in Wisconsin.

Circa 1907, $4-6

Tugboats would guide log rafts along the Mississippi at one- to two-miles-per-hour.

Cancelled 1908, $6-8

Logging on the Kentucky River.

Men work a log float from the deck of a flatboat. River loggers worked to recover any logs hung up on banks, sand bars, or islands and return them to the river channel.

Circa 1920s, $4-6

A contemporary photochrome card depicts "one of the last of the great river drives in North America" on the Clearwater River in Idaho. The annual event saw millions of feet of logs guided down the rushing stream and seventy miles to the Potlatch Forests. Begun in 1928, the drives continued until 1971.

Circa 1960s, $4-6

A tugboat pulls thirty-two-foot logs across Lake Chatcolet in Idaho. The log drives were hard on the rivers, leaving behind pieces of wood, bark and sediment.

Circa 1960s, $4-6

Logs are hauled out of the water and stacked to be processed in a mill in Missoula, Montana.

Circa 1920s, $4-6

A typical lumber camp and sawmill in the early twentieth century.

Dated 1913, $4-6

The log jams eroded the banks of rivers and deposited a great deal of debris there. Here, logs are shown in an action shot as they first meet the water in Tacoma, Washington.

Circa 1960s, $4-6

A tugboat hauls a huge boom across Puget Sound in Washington.

Circa 1970s, $4-6

A log drive on Shasta Lake is overlooked by Mt. Shasta in California. *Photo by A. Devaney, Inc.*

Circa 1970s, $4-6

A spectacular view of a fleet of log rafts near Tacoma waits to be driven to a mill. Such fleets could get as long as a half mile. To the left are tidal flats, and in the background, a mothballed Navy fleet and Mt. Rainier can be seen.

Circa 1960s, $4-6

The huge booms, with this parallel-log arrangement, looked like long highways upon the water.

Circa 1920s, $4-6

Loggers built chutes to transport logs quickly from nearby woods to rivers. Obviously you would not want to be in the path of an oncoming log!

Cancelled 1910, $4-6

From the back: "First in world production of lumber. Trees are cut from Pacific Northwest Forests and towed down the many streams and rivers to huge saw mills."

Circa 1950s, $4-6

A view from the other angle—logs coming off the tilted train beds.

Circa 1910, $4-6

Dumping Logs from train into river, Washington

670. Dumping Logs into the Water from Flat Cars.

As technology improved, the train played an important role. Here, angled tracks allow workers to more easily roll the cargo off the flatbed and into the water channel.

Circa 1910, $4-5

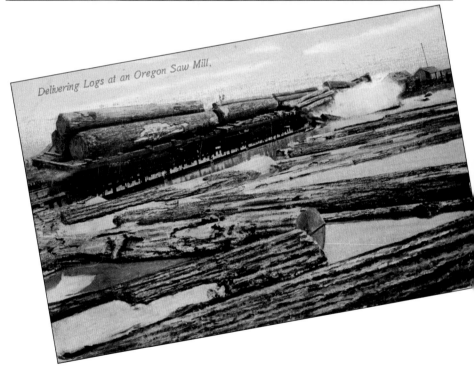

Delivering Logs at an Oregon Saw Mill.

Workers look miniscule behind the great fir logs they are working to dislodge into the water.

Cancelled 1912, $4-5

174. A LOG BOOM READY FOR THE MILL, WASHINGTON.

A boom floating in storage at a lumber mill in Washington State.

Circa 1920s, $4-6

Large booms passed through narrow waterways very slowly and carefully, like barges going through canals. Here, a giant boom negotiates Deception Pass in Washington.

Circa 1940s, $4-6

2760 – Log Tow, Snohomish River, Washington.

Cradle for Ocean-Going Log Raft, Puget Sound, Washington.

Early lumbermen believed that rafting improved lumber by drawing the sap from the wood. Although sap may have been drawn from the logs during the rafting drive, they were covered with dirt and grit. As a result, pullers, the men who pulled the logs from the water, had to wash the logs before sending them on to the mill.

Circa 1910, $4-5

In the Pacific Northwest, loggers built cradles—a wooden frame that enabled the men to pack logs together into a sturdy raft—to haul the lumber in the ocean waters off the northwest coastline.

Circa 1910, $6-8

OCEAN-GOING LOG RAFT

An ocean-going raft heads into the Pacific, off the shores of Washington.

Circa 1905, $5-7

A crib, commonly known as a deep-sea raft, was basically a gigantic bundle of logs held together by ropes or chains. Here, one raft in Oregon is shown full, another empty and awaiting its load.

Circa 1920s, $4-6

269. SEA-GOING LOG RAFT, 8,000,000 FEET OF TIMBER, OREGON.

922 — LOG RAFT OF 8,000,000 FEET OF TIMBER

Some cribs were several hundred feet long containing about eight million feet of timber.

Circa 1910, $4-6

71

The Columbia River, which is part of the Washington-Oregon border, was large enough for the deep-sea rafts to be towed up river.

Circa 1905, $4-5

Log Raft on the Columbia River.

Bridges were another problem for river floats— they could lead to log jams. Here, a raft is guided through the Mission Street Bridge in Portland. Onlookers marvel from the top.

Cancelled 1911, $4-6

LOG RAFT, MORRISON ST. BRIDGE, PORTLAND, OREGON

A painting published by British postcard company Raphael Tuck & Sons depicts a raft and river logger as they approach the Willamette River bridge.

Cancelled 1910, $4-6

Booms are disassembled and positioned for production in the mill.

Circa 1910, $5-7

The bank of this log pond is a compact skid road built perpendicular to the water, giving the ground an even surface for the quick dumping of logs.

Cancelled 1959, $4-6

Logging Operations

A scaler's job was to estimate the length of each log when the booms reached the lumber mills.

Dated 1968, $4-6

Lumber Scene.

Logs were marked with brands as well, and sorters had to know hundreds of brands. Sometimes, they had to wade into icy waters to sort logs.

Circa 1905, $8-10

When they arrived at the mills,
the logs were sorted by size,
shape, and tree type
by men known as sorters.

Circa 1914, $4-6

The men who worked on river floats were known collectively as "riverhogs," and sported such colorful job titles as drivers, jammers, sackers, and tally boys.

Cancelled 1913, $4-6

As the forests of the Pacific Coast fell to the logging industry, the industry moved north, from California through Oregon and Washington and into Canada. Shown, is a large bag boom floating down the Nepisiquit River in New Brunswick, Canada.

Cancelled 1965, $4-6

Circular booms resemble huge lily pads as they float down the river in Northwestern Ontario, Canada.

Circa 1964, $4-6

Animals—Oxen and Horses

To drag timber from the woods and haul it to the mills required powerful animals—oxen and horses. Teams, twenty to thirty of both, were common at lumber camps before the inventions of the old donkey engine and truck. These strong beasts plodded through the woods, dragging a few logs at a time while teams hauled huge wagonloads of wood to the mills. Eventually, trains and trucks replaced them.

Getting out Logs in the Green Mountains near South Royalton, Vt.

Since the early days of logging, oxen were a main mode of hauling logs from the deep woods. Here, they are shown laboring near South Royalton, Vermont.

Circa 1910, $4-6

The largest load of logs ever loaded in Vermont.

Horses pull a sled heaped with logs, as well as three men confidently sitting astride the seemingly perilous heap.

Circa 1910, $$4-6

Lumbering in the Adirondacks, N. Y.

In eastern logging centers (like this Adirondack scene shown in New York), because snow eased the transport of logs, the summer's harvest was stockpiled until winter came, and then the harvest was moved to the mills.

Circa 1910, $4-6

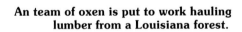

An team of oxen is put to work hauling lumber from a Louisiana forest.

Circa 1905, $4-6

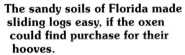

Hauling Logs to the Mill, Florida.

The sandy soils of Florida made sliding logs easy, if the oxen could find purchase for their hooves.

Circa 1910, $4-6

Ox team hauling timber.

A two-span team hauls a sizeable load of small logs on a sled. The load appears bulky, but the logs were short in size for the pair of horses to manage hauling in the snow.

Cancelled 1909, $4-6

Hauling raw material in northern Minnesota.

Cancelled 1911, $4-6

5 HAULING RAW MATERIAL, NORTHERN MINNESOTA.

"Big wheels" were devised to make dragging logs from the forests easier. This image portrays lumbering in Traverse City, Michigan.

Circa 1930s, $4-6

A team of horses pulls an empty sled back into the woods to be loaded with lumber.

Circa 1907, $4-6

Logging near Mancelona.

Lumbermen work to haul lumber out of the woods near Mancelona, Michigan.

Circa 1910, $4-6

Huge trees were sometimes referred to as "widow-makers"—trees that fell on the men who cut them down.

Cancelled 1907, $4-6

Scene at the Excelsior Plant, Coffeyville, Kans.

SIX YOKE OF OXEN HAULING 12750 FEET OF WASHINGTON FIR ON DISCOVERY BAY, WASHINGTON.

The use of oxen gave rise to another valuable worker— the bull driver. He oversaw and maintained the modes of transporting logs.

Cancelled 1913, $$-6

The men in lumber camps also had to act as blacksmiths, shoeing their oxen and horses.

Circa 1907, $4-6

LOADING TIMBER, LUMBER SCENE, WASHINGTON.

Another important job was that of the skid greaser. His responsibility was to swab the logs of the skid roads.

Circa 1910, $4-6

LOGGING IN OREGON.

A LOGGING TEAM IN OREGON

It was not uncommon for lumber camps to use teams of twenty to thirty oxen for hauling.

Circa 1905, $4-6

Logging Wheels
Knotts Berry Place
Buena Park California 6

To use the big wheels, the front ends of logs were chained to the axle and raised above the ground while the rear of the logs were left to drag on the ground.

Circa 1930s, $5-7

One of the disadvantages of using oxen was the cost of maintaining them, primarily the cost of feeding them.

Circa 1912, $5-7

Hauling Wood with Oxen Team in Santa Cruz Mountains, Calif.   129

Logging Scene. Ox Team hauling logs to the Lumber Mill.

Logging in California.

The greasing of skid roads allowed the lumber to slide much easier than dragging it across the ground.

Circa 1905, $4-6

An impressive team of ten oxen pulls a load, part of a dusty caravan helping to move mountains of California's trees to hungry mills.

Cancelled 1908, $5-7

Typical Canadian Scene. Lumbering

Two horses are hitched to a seemingly impossible load of Canadian logs.

Circa 1915, $4-6

Lumbering in the Canadian West

Men known as road monkeys built and maintained ice roads, slick enough for sleds by making grooves for the sleds' runners. Traction also had to be maintained for the animals' hooves.

Circa 1908, $4-6

Buckers cut the fallen trees
into sections so that the logs
could be hauled.

Circa 1914, $4-6

Teams of oxen and horses gather
around a primitive sawmill. Printed by
a firm in New York City, this operation may have been in New York State,
Connecticut, or the vicinity.

Cancelled 1913, $7-9

Here, a pair of oxen pulls up at an even more simplistic mill to deliver a load. The work of lumbermen and their teams of oxen and draft horses was a slow and steady one.

Circa 1930s, $4-6

Oxen stand on a skid road, the evidence of their labors behind them.

Circa 1914, $5-7

The responsibility of the bull puncher was to goad the team of oxen along to keep them moving on schedule. This real photo card was produced by Eastman's Studio in Susanville, California.

Circa 1920s, $4-6

After clearing the roads of snow, road monkeys often poured water on the roads to form ice, allowing the sleds to glide effortlessly in the winter.

Cancelled 1909, $4-6

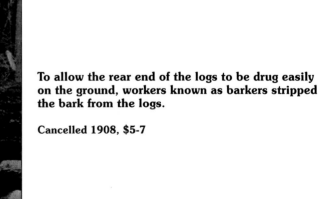

To allow the rear end of the logs to be drug easily on the ground, workers known as barkers stripped the bark from the logs.

Cancelled 1908, $5-7

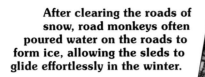

The trip to a sawmill or landing was a test of a teamster's endurance in the winter. The teamster, who drove the horse or oxen, had to sit motionless for some time on a load of logs in bitter cold.

Circa 1920s, $4-5

The teamsters' responsibilities were numerous, from repairing harnesses to nursing the cuts and bruises of their animals.

Circa 1907, $4-6

"In the Beechwoods,"

Trains

Trains enabled lumber companies to deliver wood to mills both nearby and miles away. As a result, the production of lumber increased considerably. Lumber companies scheduled daily trips to keep pace with the growing demand of wood. Long trains loaded with wood steamed and creaked across the land. Harvested lumber fueled locomotives, which could haul at least twenty-five cars full of wood.

Just as lumber camps used some of their wood for building skid roads, eventually they used wood for making railroad ties.

Circa 1907, $4-6

The locomotive innovation within the lumber industry began with placing an old donkey on a railroad car to pull timber from the forests.

Circa 1907, $5-7

Railroad tracks replaced the skid roads, although the pattern of the ties still resembled them.

Circa 1920s, $4-6

Ready for the Mills

Logging trains sped up production at sawmills, feeding the growth of the processing facilities.

Circa 1907, $5-7

Three of the eight types of engines built for industrial use were used by the logging industry: the Shay, the Heisler, and the Climax (pictured here in a postcard image from a painting by Richard Ward).

Circa 1970s, $4-6

The production of lumber increased dramatically with the various mechanical innovations, such as the train, that arose during the Industrial Revolution.

Dated 1910, $5-7

Rockslides, mudslides, and fallen trees were the main enemies of the logging trains. They required significant amounts of time and manpower to clear the tracks. Here, an engine is pictured rounding a mountainous bend in Kentucky.

Copyright 1908, $4-6

Trains may have made transporting logs easier and faster, but loading them onto the railway cars was still slow and tedious. Here, men are pictured loading train beds in Mobile, Alabama.

Circa 1903, $4-6

Despite the prosperity brought about by the industrial progress of the nineteenth century, unfortunately the workers received little of it. This card illustrates how one redwood from California could fill one train.

Cancelled 1912, $4-5

A train hauls 105 cars of lumber through Potlatch, Idaho, measuring a mile end to end.

Cancelled 1910, $6-8

One convenience of hauling logs by train was that there was plenty of wood to fuel the engines.

Circa 1907, $6-8

Not only did the innovation of train transportation increase timber production, but it also added many new jobs to the industry—far more trees could be cut and processed.

Circa 1907, $6-8

93

NO. 1303. LOG TRAIN, WASHINGTON

TRAIN LOAD OF WASHINGTON FIR LOGS.

Trains also brought new dangers to the men who worked them. Flat cars that lacked air brakes were deadly; they could break free and rush uncontrolled down the tracks.

Circa 1907, $4-6

The logging industry now needed men to run the train engines, load and unload logs from the flatcars, and men to accompany the trains in case of emergencies on the tracks.

Circa 1920s, $7-10

LOGGING TRAIN ON THE WAY TO BELLINGHAM'S MILLS.

A lumber train on its way to Bellingham's Mills in Washington.

Cancelled 1909, $7-10

287. Cedar Poles from Lower Pend Oreille,
Northeastern Washington.

A load of cedar poles
passes more woodland
cargo in northeastern
Washington.

Circa 1909, $7-10

Until the twentieth century, most
bridges were made of wood. Essential
to the development of the West, such
trestles were architectural marvels.
Here, a train crosses a trestle near
Tacoma, Washington.

Circa 1950s, $4-6

"Logs To You" from The Pacific Northwest

By the beginning of the twentieth century, lumber workers were one of the largest industrial groups in the United States. Here, three workers ride the logs aboard flatbed cars; Mt. Rainier can be seen in the background.

Circa 1950s, $4-6

Even as mechanical technology improved in the early twentieth century, lumber was still a valuable resource for the military for gunstocks, barracks, landing boats, stretchers, crutches, cots, footlockers, and trucks.

Cancelled 1911, $4-6

One log stretches the length of three passenger cars in Bellingham, Washington. With the inventions of the telegraph and telephone, huge trees were needed as poles to support the wires.

Circa 1914, $4-6

A Lumber Train, Washington.

In the late 1800s, lumber trains carried redwood trees from the dense forests to the sprouting lumber mills on the West Coast.

Cancelled 1911, $4-6

Log Train, Wash.

One locomotive used on the logging trains was the thirty-ton "Grasshopper," which was able to haul at least fifteen flat cars and twenty-five log cars.

Circa 1910, $4-5

267. HAULING GIANT TIMBER, OREGON.

The men who worked on the logging trains included brakemen who manually set the brakes on each car, an important job when trains traveled over steep hills.

Circa 1920s, $4-6

In an Oregon Logging Camp

An engine loads up with cargo to carry it from an Oregon logging camp to the mill.

Dated 1910, $6-8

On the Willamette River, North of Oregon City, Oregon. The logs are being unloaded from cars for the paper mills. The Steamer N. R. Lan is transferring a cargo of paper for shipment to Eastern Markets.

Here, a train unloads its cargo into the Willamette River, north of Oregon City, to be processed into paper. The Steamer *N. R. Lan* will transfer finished paper to eastern markets.

Circa 1911, $6-8

AN OREGON LOGGING TRAIN

Hundreds of logs are piled on a loading dock awaiting transfer to a train and on to the mill.

Circa 1914, $6-8

An advertising postcard from the McCloud Lumber Company in McCloud, California, boasts of the mill's enormous square footage: The saw mills produced 100 million feet of board annually, 16 million pieces from the lath mills, 75 million feet from the planing mills, 35 million feet from the dry kilns, 10 million feet from the box factory, 155 car loads of mouldings, and more. The operation covered 700 acres.

Circa 1926, $4-6

No. 28—"Three Log" Load of Sugar Pine at the Mill Pond.

Often, logs were delivered to a sawmill's millpond. The millponds were usually heated and allowed the mill to saw wood year round. This card was produced by Union Oil Company's "Natural Color Scenes of the West." The card states that: "Bend, Oregon ... is in the heart of the great yellow pine forests of the Pacific Northwest and has two of the world's largest pine sawmills."

Circa 1945, $4-6

Sections of California Redwood make their way via train to the mill.

Circa 1905, $5-7

HAULING REDWOOD LUMBER IN CALIFORNIA

BEAUTIFUL CALIFORNIA. LOGGING TRAIN ON A ROAD OF A THOUSAND WONDERS

Brakemen moved from car to car to apply the brakes when the train traveled down steep grades or around sharp curves.

Cancelled 1911, $4-6

M. Rieder, Publ., Los Angeles. Cal Made in Germany 8042.

A Pacific Big Stick.
Length 182 ft., Diameter 9 ft 5 in.

To haul a huge California Redwood tree, the log had to be securely chained to two flat cars.

Circa 1910, $4-6

The back of this card states that "This logging scene is typical of many log trains that the tourist will see in Washington and Oregon. Logging is still the primary industry of these two states."

Circa 1960s, $6-8

Lumber Mills

As timber demand and production continued to increase, more and more lumber mills opened for business, providing jobs for many in the small rural towns. The machine age brought radical changes to the sawmills; machines now did the work of men. Mill owners realized that they could consolidate their operations by building additional facilities onsite to provide more products such as pulp and shingles. In turn, the mills grew in size and supplied even more jobs to surrounding communities.

Although the idea of the sawmill can be traced to Germany and Scandinavia, its rise in America can be attributed to the mechanical innovations of the nineteenth century. Here, one of the many mega mills that grew up along the Pacific Northwest. Ektachrome image by Bob and Ira Spring.

Circa 1950s, $4-6

An overview shot shows a milling operation in Hallowell, Maine—the mill pond where lumber was stored, and various buildings associated with processing lumber.

Circa 1907, $4-6

1231. Pacific Creosoting Co., the Largest Timber Preserving Plant in the World, located at Eagle Harbor, Puget Sound.

As logging technology developed, the facilities at sawmills grew, with areas such as the millponds, sawing areas, kilns, and the veneering operations added or enlarged to increase production at the mills. Shown is an artist's rendering created in postcard form to promote Pacific Creosoting Co., "the Largest Timber Preserving Plant in the World, located at Eagle Harbor, Puget Sound."

Circa 1907, $4-6

Mounds and mounds of logs, scrap wood, and pulp are common sights at mills. Shown are Pulp wood piles at the Oxford Paper Mills in Rumford, Maine.

Circa 1920s, $4-5

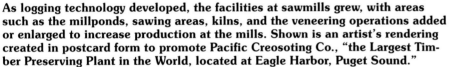

MILLINOCKET, ME.   LOG PILE OF THE G. N. P. CO., 21 MILLION LOGS.

Years ago, most mills could cut at least 10,000 feet of lumber per day. Modern mills produce at least twenty times that each day. Here, log piles feed a millpond, and a mill at the G. N. P. Co. of Millinocket, Maine. Wood piles have to be kept wet as the curing lumber creates tremendous heat, and can burst into flame.

Circa 1910, $4-6

104

As in other industries, lumber companies began consolidating operations by purchasing smaller mills and forest properties. Shown is the TMH Lumber Co., Inc. in Tappahannock, Virginia.

Circa 1960s, $4-6

540—CHAMPION PAPER AND FIBER COMPANY, CANTON, N. C.

Shown is wood stockpiled for processing in the Champion Paper and Fiber Company of Canton, North Carolina.

Circa 1940s, $4-6

Lumber Laden Vessel, North Tonawanda, N. Y.

A ship takes on a load of finished lumber at North Tonawanda, New York.

Circa 1920s, $5-7

Four and five mast schooners comprised many of the coastal fleets before steams engines replaced them. Steam schooners could haul more lumber quicker than the sailing ships.

Cancelled 1909, $5-7

Shipping Lumber, Savannah, Ga.

A promotional card for Tampa, "The Cigar City" cites the Lumber Fleet's accomplishment of shipping as much lumber in 1929 as was shipped between 1920-1927!

Circa 1930, $5-7

THE LUMBER FLEET, TAMPA, FLA.　　THE CIGAR CITY　99

EASTMAN GARDINER LUMBER MILLS, LAUREL, MISS.—13

Millponds were heated by the steam plants at sawmills. By softening the wood, the mill could operate year round, providing steady business and employment. This mill was in Laurel, Mississippi.

Cancelled 1944, $4-6

By the mid 1800s, inventions such as scroll saws, cylinder saws, mortising machines, and the bed plane enabled the lumber industry to keep pace with the rising demand of wood products. This is the Big Mill at Chippewa Falls, Wisconsin.

Cancelled 1911, $4-6

BIG MILL CHIPPEWA FALLS, WIS.

2157

A hardwood sawmill and power plant in Crossett, Arkansas.

Circa 1930, $4-6

HARDWOOD SAWMILL AND POWER PLANT — CROSSETT, ARKANSAS

D-5466

2383. Log Entrance, Paine Lumber Co., Oshkosh, Wis.

The intake chutes at the Paine Lumber Company in Oshkosh, Wisconsin.

Circa 1920s, $4-6

The Virginia & Rainy Lake Co.   VIRGINIA, Minnesota.
Loading Lumber, Pile to Cars.

Lumber is loaded piece by piece for transport from the Virginia and Rainy Lake Company in Minnesota.

Circa 1920s, $4-6

No.10.

The process of making plywood has been around for thousands of years. Thin sheets of fine wood—in this case knot-free Douglas fir—are peeled from logs by a rotary lathe and glued to lower-quality wood.

Circa 1940s, $4-6

MAKING FIR PLYWOOD—P106

Giant lathes are used to peal the veneer from the fine, knot-free Douglas Fir used in the making of fir plywood. The block is rotated against a keen-cutting blade (at back) which "shaves" off the ribbon of wood usually from a tenth to a seventh of an inch thick.

These timber blocks of spruce were cut from trees ranging from 200 to 300 feet high. Spruce is used in many things, from the making of stringed musical instruments to strong structural beams for construction. Its long wood fibers are perfect for paper production, too.

Circa 1950s, $4-6

869—OREGON "TOOTHPICKS," FROM AN OREGON FOREST

29989-N

An image from the Blackwell Lumber Company in Coeur d'Alene, Idaho—the largest planing mill in the world.

Cancelled 1909, $5-7

The large saws used to cut logs had to be filed and sharpened twice a day. Shown, a log mounted on a carriage is processed through a sawmill in Oregon.

Circa 1914, $4-6

GOING TO MILL FROM AN OREGON FOREST.

Two men pose by a sixteen-foot spruce, soon to be reduced in an Oregon mill.

Circa 1920s, $4-6

Hammond Lumber Mills, showing millions of feet of floating lumber, Astoria, Oregon.

A man stands amidst the inventory in a large millpond in Astoria, Oregon.

Circa 1910, $4-5

A conveyor belt takes a great log into the mill, pausing so two workers can pose for an image.

Cancelled 1913, $5-7

1469- Log entering Sawmill. Oregon.
Copyright by Weister.

111

1506. Bunker Hill & Sullivan Buildings, Kellogg, Idaho.

The Bunker Hill and Sullivan Buildings in Kellogg, Idaho—a massive mill operation.

Circa 1910, $4-5

Union Oil Company produced this card as part of its "Natural Color Scenes of the West" series. On the back they state that "Lumbering in the Northwest serves as a general stimulant to other lines of business through its need of equipment and supplies. Millions of board feet are cut every year from the nation's finest stands of timber."

Circa 1947, $4-6

A bird's-eye view of a sawmill in the Northwest shows its typical parts: the burner, the mill facility, the millpond, lumberyard, and train dock.

Circa 1940s, $4-6

The largest waste material at a lumber mill was sawdust, and what wasn't used as fuel at the mill was sold locally for fuel and farming. *Photo by L. L. Perkins.*

Circa 1960s, $4-6

Coastal lumber mills took advantage of their locations by selling lumber to local shipbuilders. In addition, they used the wood from their mills to build wharves and shipping facilities.

Circa 1940s, $4-6

1038—Port Blakeley Lumber Mill, Washington.
Copyright, 1904, by Lowman & Hanford S. & P. Co., Seattle.

Loading
Lumber
Port Blake
Washingto

The Port Blakely Mill Company in Washington was one of the world's largest sawmills. It was engaged in shipbuilding and the shipping of lumber throughout the world.

Copyright 1904, $4-6

By the early twentieth century, three mills had been built and owned by a former sea captain from Nova Scotia, Captain William Renton, on Brainbridge Island at Blakely Harbor, Washington.

Circa 1910, $4-6

Loading Lumber at one of the Large Puget Sound Mills,   Tacoma, Wash.

By 1878, at least sixteen logging camps arose along the Skagit River, near Puget Sound. Here, lumber is loaded on sailing vessels at Tacoma.

Circa 1910, $4-6

174. A LOG BOOM READY FOR THE MILL, WASHINGTON.

Log booms usually towed timber to the tidewater mills, like this one in Washington State.

Circa 1915, $4-6

1436- A Washington Saw Mill

Even as they floated in the water, the logs still appeared gigantic to the boom men who hopped around on them.

Circa 1920s, $4-6

Occasionally, lumbermen bought timberland and built their own mills. However, these arrangements often left the lumbermen/owners cash-strapped, forcing them to log quickly, pay low wages, and sell at cost.

Circa 1960s, $4-6

Large drive of shingle bolts on Stillaguamish River, Arlington, Wash.

Arlington, Washington was once known as the "Shingle Capital of the World" because of its many shingle mills like this one.

Circa 1914, $4-6

Named "a most interesting stop on the Redwood Highway," The Pacific Lumber Company (PALCO) was one of the largest redwood lumber mills located in the town of Scotia, in northern California.

Cancelled 1936, $4-6

PACIFIC LUMBER COMPANY, REDWOOD MILL

PHOTO BY CALIFORNIANS, INC.   REDWOOD HIGHWAY, SCOTIA, CALIFORNIA   6A-H125

116

As the timber industry exploded in the Northwest during second half of the nineteenth century, San Pedro became a major port in the shipping of lumber on the Pacific Coast.

Cancelled 1909, $5-8

A lumberyard in Fellows, California, pre-World War II.

Circa 1914, $4-5

At one point in the twentieth century, lumber was the main industry in Redding, California.

Circa 1905, $4-6

A sawmill in California displaying two sections of a log over ten feet in diameter.

Circa 1950s, $4-6

The crane was a huge innovation for the lumber industry in the early twentieth century for the ease with which it loaded and unloaded lumber.

Circa 1930, $4-6

A Lumberman's Workshop

An artist captures the atmosphere of a California sawmill—a lumberman wearing traditional suspenders, a huge band saw, a tree stump as a workbench, and colossal logs to work on.

Circa 1910, $4-6

Lumber was still the most vital material needed during the second World War.

Circa 1940s, $4-6

119

Sluicing Lumber at Bangor Dam, Bangor, Me.

In the nineteenth century, Bangor, Maine, was a hub for the timber business in the Northeast.

Cancelled 1909, $4-6

Lumber companies lured workers to remote mills by building stores, houses, and small businesses. By creating and essentially owning their small communities, the lumber companies contributed to the end of "company towns"—towns that relied on a company as the main employer.

Circa 1940s, $5-7

Trucks

After World War II, the truck became the primary manner of lumber transportation. Small trucks ground up and down the old logging roads previously deemed inaccessible to vehicles. They hauled timber to waiting tractor-trailers that carried the wood to the mills. As a result, companies could exploit more remote areas of the forest.

Logging Truck in North Idaho

Photo: Bill Hawkins

A small logging truck carrying timber in the south in the early 60s.

Circa 1978, $4-6

A truck works to extract big timber from the woods of North Idaho.

Cancelled 1980, $4-6

Douglas Fir Logs - Washington

When the diesel engine became more common in the large American trucks after World War II, the tractor-trailer became an important part of the logging industry. Here, a big rig hauls Douglas fir from Washington's Cascade Mountains region.

Circa 1980s, $4-6

MT. ST. HELENS

On May 18, 1980, Mt. St. Helens in Washington State erupted, devastating some 150,000 acres of private, state, and federal forests. The Weyerhaueser Company salvaged more than 850 million board feet of timber from within the blast zone, according to the back of this card.

Circa 1980s, $4-6

A convoy of logging trucks hauls huge fir trees in western Washington State.

Circa 1930s, $5-7

GIANT FIR LOGS - WESTERN WASHINGTON

180   GIANT LOG EN ROUTE TO MILL, 9 FT. 11 INCH DIAMETER, 20 FEET LONG, 6980 BOARD FT.

6A-H3224

A card boasts of a logging haul—a 9-foot, 11-inch, 20-foot long chunk of fir tree. It was estimated to produce 6,980 board feet.

Dated 1967, $4-6

Log Truck Unloading in the Pacific Northwest

The huge logging trucks were able to travel on the old, rugged logging roads to and from the lumber mills.

Circa 1960s, $4-6

Giant fir trees are still harvested in Washington and Oregon, typically used for paper and cheap construction products such as packing crates.

Cancelled 1969, $4-6

"Five Houses" Enroute to the Sawmill

From the back: "A load of 14,000 board feet is being transported to a nearby mill or railroad."

Dated 1950s, $4-6

As the diesel engine became common in the huge logging trucks, a few gasoline engines were still used in the 1970s.

Circa 1970s, $4-6

124

In their early years of use, Peterbilt and Kenworth trucks were the primary heavy-duty trucks used by the lumber industry.

Circa 1970s, $4-6

Lumber businessman T. A. Peterson rebuilt surplus army trucks in order to haul felled trees from the deep forests to his lumber mill in Tacoma, Washington. Eventually, he founded the company, Peterbilt Motors, in 1939.

Circa 1970s, $4-6

16-469 Hauling a Forest Giant in Oregon

In 1933, Kenworth was the first truck company to have the diesel engine as a standard feature.

Circa 1940s, $5-7

125

An example of a "bridge load"—the log resembles a "bridge" from the truck to the trailer, with one end resting on each.

Circa 1950s, $4-6

Such a logging scene, where a logging truck crosses a wooden bridge, is still prevalent in the second half of the twentieth century.

Circa 1970s, $4-6

To haul the giant tree logs that would be used as telegraph/telephone poles required a longer trailer.

Cancelled 1940s, $5-7

Diesel was one-third the price of gasoline, helping lumber companies keep their transportation costs as low as possible.

Circa 1970s, $4-6

From the back: "In the forests of Northern California, where multiple use has become a way of life rather than just the dream of a few idealists. That a forest can provide both lumber for your home and recreation for your family has been proven both possible and practical."

Cancelled 1973, $4-6

A logging truck arrives at the log "dump" in Oregon, with a storage pile in the background.

Circa 1970s, $4-6

A "Welcome to Oregon" postcard stars the state's primary industry—logging.

Copyright 1969, $6-10

127

Modern logging crews still contain rugged men of various ages and backgrounds, and the accomplishment of wrestling a giant tree to the mill still merits a photo op.

Circa 1970s, $4-6

A superb depiction of an early logging truck hauling logs from the immense woods of the Northwest.

Circa 1930s, $4-6

The Redwood Highway, the nickname of Highway 101, runs through northern California and is known for its spectacular views of the redwood forests.

Circa 1960, $5-7

Hauling logs to the mills by truck is known as secondary transport.

Circa 1930s, $5-7

From the back: "Motorists throughout Northern California will find trucks such as these a familiar sight as huge timbers are hauled to the sawmills."

Circa 1960s, $4-6

A logging truck could haul up to 9,900 pounds of wood—a little less than five tons.

Circa 1970s, $4-6

Most trucks were designed to travel off-highway on the rocky logging roads and for short-distances on the highway.

Circa 1970s, $4-6

A picture containing the two primary modes of modern lumber transportation: the logging truck and the railroad flatcar. From the back: "A load of 14,000 board feet (one log) enough to build 5 complete 6 room houses."

Circa 1960s, $4-6

The huge trucks contained powerful diesel engines to enable them to negotiate the steep, rugged territory of the northwest woods.

Circa 1980s, $4-6

By the end of the Vietnam War, the military began using a lighter truck in its engineering units, and the surplus trucks were once again acquired by the logging industry.

Cancelled 1972, $4-6

Logging trucks hauled either huge log sections or a load of smaller logs, depending on the particular wood order. *Photo by Gan Nilsen.*

Circa 1965, $4-6

Contemporary Images

Although fewer mills operate today, modern mills provide scenes of productivity from their huge lumberyards to wood pulp piles. The brilliant chrome images produced on postcards during the second half of the twentieth century illustrate the vitality of the mill, showing the size and scope of the operation. Massive machinery, like loaders and cranes, are common at today's thriving mills. Lumbermen no longer swing the axe or thrust saws, as the chainsaw is now the tool of choice.

Often forgotten in the history of logging—it was at Sutter's Mill where gold was discovered in California in 1848. The treasure appeared in the ditch that carried water from the American River to power the sawmill.

Circa 1970s, $2-4

From the back:
"Thousands of cords of pulp
wood are delivered here by truck and rail,
then piled in a mountainous mass waiting to be ground
for use in the hungry paper machines at the massive Rumford, Maine
mill site."

Circa 1960s, $2-4

The largest paper pulp mills
in New England are found in
Rumford, Maine.

Circa 1960s, $2-4

A huge pile of pulpwood waiting to be ground and made
into paper in Maine.

Circa 1960s, $2-4

The mountains in Pennsylvania
provided a once thriving lumber
industry, and are still home to a
number of sawmills.

Circa 1960s, $2-4

From the back: "Pulpwood is rafted across Lake Superior from points in Minnesota and Canada, to Ashland (Wisconsin) where it is reshipped by rail to mills at Wisconsin Rapids."

Circa 1960s, $2-4

Located in northern New Hampshire, the town of Berlin was once the center of the region's lumber and paper mills. Sadly, in March of 2006, the last paper mill in Berlin closed. *Photo by Don Sieburg.*

Circa 1960s, $2-4

The familiar sight of a log boom on the St. Mary's River near the U. S.-Canadian border on the upper Michigan peninsula in a photo by Lucy Gridley.

Copyright 1963, $2-4

The R. F. Learned and Son, Inc. Band Sawmill had been in continuous operation since 1825 at Natchez-Under-the Hill, Mississippi.

Circa 1960s, $2-4

A large pulpwood storage facility in Minnesota—logs from all over the state were shipped here, and then to a mill a half mile away. *Photo by Bob Glander.*

Circa 1960s, $2-4

A lumber mill and its millpond in Big Sky country, Montana.

Circa 1960s, $2-4

A view of the largest lumber mill in the world, owned by Weyerhaeuser Timber Company Plant, and in operation since 1900 in Washington State.

Circa 1960s, $2-4

A millpond in the Pacific Northwest, photographed by Lyman B. Owen.

Circa 1960s, $2-4

135

After the end of World War II, the increased demand for wood products created a new movement of harvesting second-growth timber, which led to an explosion in tree farming.

Circa 1970s, $2-4

Oregon Timber Industry

From the back: "Harris Pine Mills. This nation-wide industry, with headquarters at Pendleton, Oregon, conducts comprehensive operations from growing its own trees to manufacturing and marketing fine furniture. Its sawmills and factories cut and process annually enough lumber to reach around the world."

Circa 1960s, $2-4

Oregon Sawmill

By the 1960s, more than one-fifth of the U. S. timber supply came from the forests of Oregon.

Circa 1960s, $2-4

At one time, southern Oregon provided one quarter of the Sugar Pine used for lumber.

Circa 1970s, $2-4

From the back: "The Port Angeles, Washington division of Crown Zellerbach Corporation produces quality newsprint for daily and weekly newspapers throughout the West. The mill is located on Ediz Hook beside the Strait of Juan de Fuca. Wood to supply the pulp mill is grown on Olympic Peninsula tree farms."

Circa 1960s, $2-4

In the early 1970s, forest conservation legislation became a bigger nemesis than wildfires to the lumber industry.

Circa 1961, $2-4

Tree farming was introduced in Washington State in 1941 to regenerate forest production.

Cancelled 1967, $2-4

On January 30, 1964, the lumber barge George Olson ran aground near Cape Disappointment along the southern coast of Washington State. Three and a half million feet of lumber came ashore.

Circa 1965, $2-4

Orderly stacks of lumber, the end product, await shipment from the sawmill.

Circa 1960s, $2-4

The city of Coos Bay, Oregon has been involved with shipping, shipbuilding, and other wood products since the mid 1800s.

Circa 1967, $2-4

The Pacific Northwest coasts provide superb shipping and loading facilities for foreign and domestic markets.

Circa 1970s, $2-4

From the back: "This mammoth loader, nicknamed the 'Elephant,' is shown here lifting the entire load of this huge logging truck with ease, then scurrying away to the mill nearby where it soon will be lumber."

Circa 1960s, $2-4

The redwood forests in northern California gave rise to many sawmills, such as this one in Redwood Empire.

Cancelled 1958, $2-4

A redwood lumber mill located in Scotia, California—the last true company town in America. *Photo by Mike Roberts.*

Circa 1960s, $2-4

Scotia was founded as a logging camp in 1883, and originally named Forestville. *Photo by Mike Roberts.*

Circa 1960s, $2-4

From the back: "These loggers climbed into this undercut to emphasize the enormity of the trunk of this giant California Redwood. These men are each over six feet tall. The diameter of the tree is 13-1/2 feet.

Circa 1960s, $2-4

Developed in the late 1920s, the gas-powered chainsaw is now the tool of choice for modern lumbermen. *Photo by Joseph Scaylea.*

Circa 1960s, $2-4

Modern lumber mills included barkers that strip the bark from the log in order to make the most of each log harvested. Here, logs await treatment at the Union Lumber Company in Fort Bragg, California.

Cancelled 1968, $2-4

Pickering Lumber Mill and Pond at Standard, California ... "is the principal industry left in this once famous Mother Lode Country of the old gold rush days."

Circa 1960s, $2-4

A mill in Flagstaff, Arizona, processes timber from the largest ponderosa pine forest in the world.

Circa 1950s, $2-4

A steamship hauls wood in the Kuskokwim River in western Alaska. *Photo by Don Horter.*

Circa 1960s, $2-4

143

Bibliography

Andrews, Ralph W. Glory Days of Logging; Action in the Big Woods, British Columbia to California. Atglen: Schiffer Publishing, 1994.

——. Redwood Classic. West Chester: Schiffer Publishing, 1985.

——. This Was Logging. West Chester: Schiffer Publishing, 1984.

——. This Was Sawmilling. Atglen: Schiffer Publishing, 1994.

——. Timber: Toil and Trouble in the Big Woods. Atglen: Schiffer Publishing, 1984.

Lillard, Richard G. The Great Forest. New York: A. A. Knopf, 1947.

Monte, Mike. Cut & Run Loggin' Off the Big Woods. Atglen, Schiffer Publishing, 2002.

Website Resources

www.Tunneltree.com
www.theculturedtraveler.com/Parks/Archives/Redwood
www.americanparknetwork.com